BEI GRIN MACHT SICH IHR WISSEN BEZAHLT

- Wir veröffentlichen Ihre Hausarbeit, Bachelor- und Masterarbeit

- Ihr eigenes eBook und Buch - weltweit in allen wichtigen Shops

- Verdienen Sie an jedem Verkauf

Jetzt bei www.GRIN.com hochladen und kostenlos publizieren

Bibliografische Information der Deutschen Nationalbibliothek:

Die Deutsche Bibliothek verzeichnet diese Publikation in der Deutschen National-
bibliografie; detaillierte bibliografische Daten sind im Internet über http://dnb.d-
nb.de/ abrufbar.

Impressum:

Copyright © 2016 GRIN Verlag, Open Publishing GmbH
Druck und Bindung: Books on Demand GmbH, Norderstedt Germany
ISBN: 9783668264786

Dieses Buch bei GRIN:

http://www.grin.com/de/e-book/337029/the-origin-of-biological-complex-gear

Michael Dienst

The Origin Of Biological Complex Gear

Design Intent regarding Surfboard fins with "Intelligent Mechanics, i-mech"

GRIN Verlag

THE ORIGIN OF BIOLOGICAL COMPLEX GEAR

Design Intent regarding Surfboard fins with "Intelligent Mechanics, i-mech"

Michael Dienst
BIONIC RESEARCH UNIT BERLIN
Summer 2016

The artificial fin system CARPO

Surfing (*hawaiian: he'e nalu, engl. surfing*) exerted on coastal waves and is in a sliding motion on the water surface. The typical for surfboards and differently designed fins at the tail stabilize the board in drive and enable steering movements in the shaft. Finns are as guidance and control wing for lateral plane of surfboards and fluidic properties and the specific selection by the surfer that are critical to performance, maneuverability and style of surfing. Profile flexible, deflectable or hingedly executed designs for surfboard fins are not prior art. The specific requirements in driving and maneuvering, make flexible guidance and control systems wings desirable with non-symmetrical airfoils. The artificial fin system CARPO is inspired by the fluid-structure interaction of elastic-moving dolphin hand. The "hands" of vertebrates form of kinematic structures that occupy adaptively and autonomously an advantageous shape under mechanical stress. This passive exercise Deformation interaction of the metacarpal (lat.: metacarpal) requires no cognitive effort of the essence. We call stress Adaptive Designs "intelligent mechanics, i-mech". The metacarpals system of dolphins is a spatially effective linkage, lead in the flow forces on a load-adaptive change in shape of the structure. The sensible semantics of functional elements makes the ordering and the kinematic principle of Biosystems visible. CARPO FIN is a first fundamental transfer of biological shaping principle "carpus" in a flow-adaptive, artificial surfboard fin. When maneuvering the CARPO FIN structure differs slightly the cross flow force (canting). Kinematic (i-mech) enters the airfoil in a spatial deformation state (adaptation) and an asymmetric profile contour is formed. Theoretical flow investigations suggest, that the now effective flow at airfoil the lift force (lift) of the fin increases and also the harmful stall (Separation) out delay time. The flow resistance (drag) of the deformed fin is less than that of symmetrical contour in off condition (stall); agility of the shear force generating increased momentum during maneuvering.

Hands

scaled from a technical, systemic view and reported our hand, in particular the system of metacarpals (metacarpal) "transmission elements" according to the above-described bending and spreading principle in a spatial arrangement. This arrangement is very complex. The carpometacarpal joints (connecting the distal hand root bones with the second to fifth metacarpals, (Articulationes carpometacarpales II-V) are not directly associated with the wrist in humans. The metacarpal bones of the vertebrate hand are the Design Intent for the development project CARPO Fin. As these supposed transmission elements are arranged and how they interact with each other, initially remains once hidden under the skin of my hand. Welfare is a man-made no obvious bionic model of a flow body. However, on an abstract level, we find a design principle of the two-legged land creature man reveals the kinematic nature of the metacarpal in a special form and interpretation (upper extremity). Let the apple in an "unsuspecting" hand drop, we see immediately that this transmission system can run without cognitive feedback (meaning, nerves and brain), and executes a wise collective movement. The loaded by the weight of an apple, the hand seems passive and adaptive to grip the object. It is an embodied prudence, a characteristic that has acquired and retained the vertebrate skeleton during the millions of years of biological evolution. So I dare - to claim that it is in my, the human hand to a system that clearly covers to load adaptive elements having "intelligent mechanism" - in the absence of an inanimate hand as evidence leading reference system.

Maybe we try the issue of intelligent mechanics better in biological systems (beings) to answer whose limbs professionally involved fluidic guidance, control and maneuvering drive tasks or have employed. We turn therefore to the dolphin hand, a wonderful example of "intelligent mechanics, i-mech" and as a motivation for the development of innovative, future-proof surfboard fins. Consider the "essence of CARPO Fin".

Whales

A large number of beings who are living in aquatic vertebrates. Over the biological evolution, some terrestrial mammals have developed an aquatic life form. The Ambulocetus ("running whale") about a genus first whales (Cetacea)

Michael Dienst, BIONIC RESEARCH UNIT Berlin, Summer 2016

from the time of the early Eocene (about 50 million years ago). Ambulocetus could run both swim. At his anatomy that suggests this amphibious lifestyle, paleontologists show today that have whales (probably rather small) terrestrial mammals developed. The pelvic girdle which is now in the further evolutionary development the seas peopling mammals receded gradually. The limbs formed around and adapted themselves to their functions of the new, the aquatic life. Arms and hands were transformed into fins (flippers) the entire body of cetaceans has been streamlined, making new body parts, such as the dorsal fin. Today we see only a selectively decrypted segment of evolutionary development and the identification of targets of recent research aims in rare cases with those of oriented applications bionics. Nevertheless, some basic instructions are extractable. The entire structural morphology of modern whales and cetaceans is a constant reference to the adapted to life on land ancestors. So the hands of modern dolphins are not only used for maneuvering, but also serve as an important tactile organs in the social and sexual relations with other dolphins. The dolphin hand out the extreme example of the functional differentiation of a fluid-mechanically effective guidance, control and drive the wing from the already extremely complex movement, which in the course of biological evolution an aquatic life-form (re) adopted. Since the preparations of the hands of dolphins and other cetaceans in most collections serve purposes other than the analysis of movements and the presentation of their complex kinematics, the extraction of the kinematic solution principle of biological transmission of interaction affection his joint elements and comes the development of a technical interpretation of the biological principle a certain importance. Prior to the design of kinematic Surfboard fins with intelligent mechanics, i-mech, functional, kinematic solution here principles of the envisaged structure are discussed. The solution principle is indeed an important work product of the early phase of industrial product development.

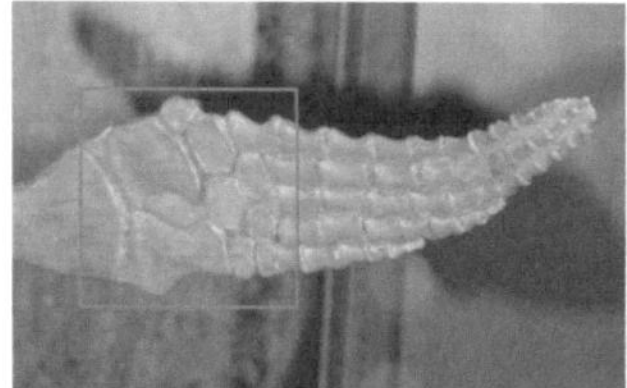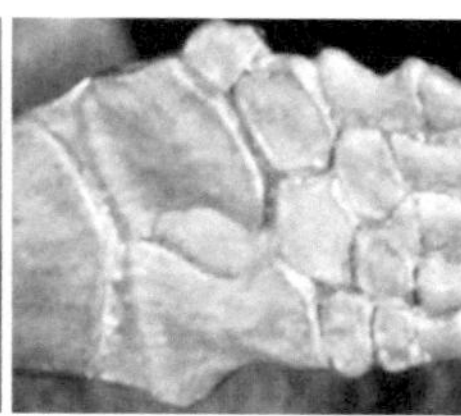

Illustration5: and Illustration 6: Preparation of a "modern" dolphin hand. Natural History Museum Berlin, Mi. Dienst 2013.

Evolution of fins

The description of the origin and evolution of the vertebrates (vertebrates) is not appropriate, gaps and has led in the past, experts repeatedly heated discussions. New fossils that phylogenetic analysis and improved methods lead to developmental models with which bodily functions - and in our case interested body kinematics - can be made understandable. An important element in the understanding of the paired extremities, especially the "handed", fluid-mechanically effective fins is their semantic positioning and arrangement in the overall skeleton of vertebrates. The body skeleton has to store the function to protect the intestines various minerals, to form a series of solid articulated (skeletal) elements and give the body a certain stiffness. The central organizational system, the spine is older than any other part of the post-cranial skeleton except the notochord, is the original, mesodermal, internal axial skeleton of all chordates and for this the eponymous feature.

The phylogeny of limb development in vertebrates is not completely. There is evidence that the original (aquatic) vertebrate has developed starting a seam of lateral fins pairs at the gills to the rear of the body. Fossil evidence suggests that modern fish have adopted the concept of phylogenetic fin base their ancestors. Experiments for induction of extremities of the amphibians larva also point out that the formation of limbs wherever (the continuous) fin folds are postulated.

The (paired) dorsal and anal fins serve to stabilize the system in service and prevent the body rotates around the vertical axis (yaw) and the longitudinal axis (roll). Probably the original (morphological and constructive) state that each fin but was rod-like support initially segmented "radials" within the body contour by an arrangement. The segmental arrangement was lost during evolution. The earliest paired fins have a broad base, so that the proximal part (towards the body) facing outwards was often much larger than the distal and basal (from the base) is called. The visible membrane of the fin was protected from dermal scales. The fins of more advanced fish were internally supported by a number slender flowed rays. The rays of the cartilage and the bone fish differ from each other and are derived from scales. Later di base of the fin was reduced, which improved their mobility. The basals widened, their number was included low and in the edge of the fin. The fins of the recent fish consist of a membrane support surface (webbing), which is stabilized by non-isotrope fin rays. In the musculature the fin rays are anchored with fin ray beams. The morphological structure of the fin represents a balanced combination of stiffness and flexibility, allowing the animal a finely tuned hydrodynamic interaction with the environment. The rays of bony fish are by sting rays (hard) and member chains (soft) distinction. Hard jets are unstructured, mostly smooth pieces of bone, soft rays consist of two halves fused together.

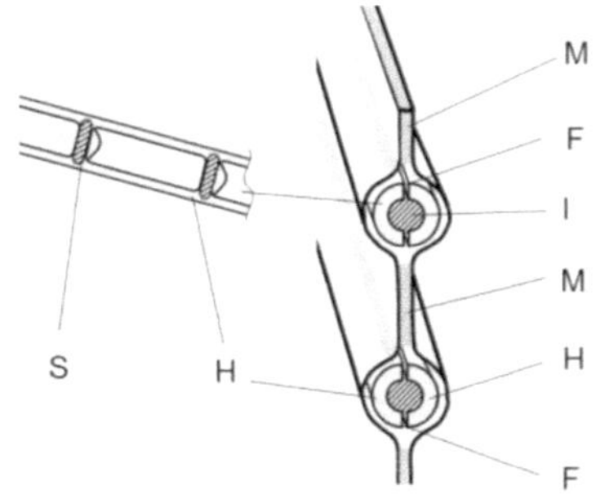

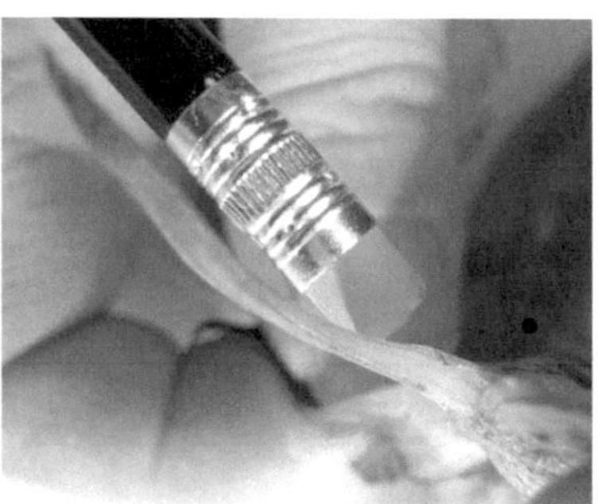

Illustration 5 and Illustration 6:
Fin membrane with fin rays (schematic illustration) web S, half-tube H, membrane M, F and Fugue Inlet I. Right: The non-orthodox motion conduct a mackerel fin.

The rays of the fish's fin as two at a certain distance linked by bridges S, structured half tubes H You have to imagine. The half-tube system has a Inlet I, filling the space between half tubes and bars. The membrane M coats the half tubes that can slide on each other due to joint F; schematic sketch in Fig.5. Fish fin rays are bilateral structures. The two halves of each beam can slide past each other. The displacement movement is in response to external loads and / or when the bases of the fin rays system, the fin muscles are moved to the root of the fin rays. The surrounding membrane (webbing) forms pockets which encase the half-tubes of fin rays and add them to a compact quasi-round material and radially stabilize. At the webs S, the two half-tubes are mechanically coupled to one another. Membrane with membrane bags and tubes systems, respectively webbing and fin rays form a three-dimensional airfoil with anisotropic properties.
The fluid-structure interaction in momentum exchange with the fluid through the membrane support surface of the fish's fin can be productive or generative.

In a productive interaction fins membrane works as a motor wing and couples energy from the flow in the membrane a. In a generative fluid-structure interaction, the fin membrane couples energy from the structure in the fluid one. Production and generation can take place in a time-locally interlaced.

Unlike in the art, where the energy and information exchange at engine and wings can be relatively clearly identified and assigned to biological wing structures provide a complex, for feedback and adaptation enabled multifunctional systems. These are optimized and able to their catching fluidic environment creatively interact with them and to condition for their transport and mobility issues so that the creatures the timing of his body movement so that the gene graced swirl the found in its surrounding structure advantageously supplements. Here have periodicity, frequency, phase and direction of rotation of the significant of a flow transported to a fin membrane eddy formations affect the quality of the fluid-structure interaction with the fin membrane.

Basically, fish are able with the fin muscles at the root of the fin rays actively control the curvature of each fin ray and thus to shape the curvature of the entire membrane in a very complex way. The decisive kinematic property of the rays penetrated fins skin membrane, a so-called "non-Orthodox stress strain interaction (NOB)" trainees-lead, based on the connected in a regular report by webs and radially bound by the webbing half tubal systems. Under a horizontal the main axis of the system and thus acting perpendicular to the tube systems distributed load cause the fin rays, an elastic concave deformation of the curvature of the loading direction is opposite. Accustomed intuitively assign a so simple beam load a convex dodge, this concave stress deformation appears to us regime paradoxical at first glance. If the momentum exchange at the membrane surface of the fin very large, the biological system behaves bending flexibility, pliability elastically and can yield a non-axial flow. the apply and shape-change- interaction correlated with the direction of working force in terms of a conventional load deformation regimes. Self conventionally deformed components so behave mechanically "orthodox". In normal operation (in technical terms, "off interpretation the flow component") shows the fish's fin a mechanically nonorthodox (NOB), indeed paradoxical deformation behavior: the one of the force initiation direction counteracting deformation realize paradoxical apply valve strain- interactions. The knowledge about the anatomical design, the functions and particularly over the impingement motion behavior of the fish's fin are still incomplete.

Michael Dienst, BIONIC RESEARCH UNIT Berlin, Summer 2016

From water to shore.
Today we know that tetrapods have developed in or near the fresh water; with their strong fins they could find food on land and aquatic escape enemies. Fossil amphibians similar (panderichthyider) fish and fish-like (ichthyostegider) Amphibians traditional interpretation models for the development of fins towards a suitable for life on the mainland leg. The gradual transformation of fins into limbs demanded a morphological reconstruction, the loss of fin rays and the formation of fingers. The legs of tetrapods have a variety of complicated functions. Raising the body demanded a lateral alignment of the (early) leg. The transition from water to the land is the subject of the relevant literature. The general public has been the history of the shape change by the Fund (1986) of "Tiktaalik roseae" actually already a wrist (wrist) and had the wonderful book "Your inner Fish" from the American paleontologists Neil Shubin accessible. The bones of the wrist make the carpus. The different skeletal pattern of tetrapod hands vary the original pattern by omitting and fusing. These operations are derived from studies of embryo development beings. The skeleton of the fluidized-tier hands formed from cartilaginous elements within the developing limb bud.

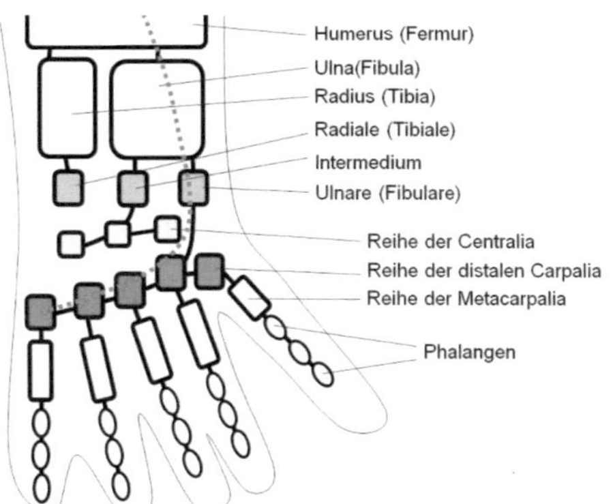

Illustration 7: Organization and principle of development of the limb skeleton of tetrapods. Marked ontogenetic development corridor dotted: Schematically by Shubin.

The elements are formed in an order (line dashed in blue) from proximal to distal from precursor elements that follow the path that is represented by black lines. Important in this context is the (relatively new) Recognizing that the fingers of the vertebrate hand did not result from a conversion postaxial radials (the evolutionary preceding fish fins). Rather fingers are a reinvention of tetrapods. It is not entirely clear why we (tetrapods) have five fingers on each hand. The early amphibians varied the basic pattern and experimented with more than five fingers. However, it is no lineage leading known to modern animals, which do not represent the concept of five-fingered hands.

The skeleton of plans (foot and) Carpal of vertebrates vary a common pattern; Change in shape in particular by scaling and reduction. In modern birds (and their alleged ancestors, the dinosaurs) are the variations of the carpal bones the most advanced.

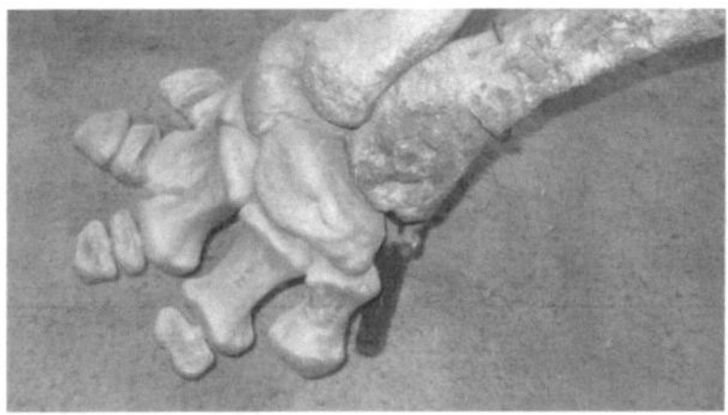

Illustration 8: Preparation of a dinosaur foot (Kentrosaurus). Natural History Museum Berlin, Mi. Dienst 2013.

The sketch Fig.7: shows schematically the principle of development of the limb skeleton of tetrapods, Fig.8 a part of the hind limb of a *Kentrosaurus*. If the pattern of the vertebrate hand (~ feet) yet each fingers (~ toe) explicitly operated from the series of distal carpal bones of the metacarpal, just put the foot ulna (fibula) and radius (tibia, top and center right, figure 8) few degenerated to remaining metacarpal elements. The figure Fig.9 shows the hand of a recent semi aquatic life (sea lions) and makes the particular problem in the preparation of the metacarpal system placental mammals clearly: in the permanent collections of museums triggers the display of complex cartilage tendon-bone collectives rule on technical limitations. We see that the ulna and radius (top and middle left, Fig.9) are slim and find - very similar to the human hand - a highly-differentiated kinematic connection to the metacarpal-system; the five movable fingers have three segments.

At first glance, the arms and hands of humans seem little related to the pectoral fins of fish. But here appears a variation of the model base for the generals Verde Roast hand. The wrist of the human described by the morphologists schematically as a toothed hinge (articulatio ginglymus). Because of its geographic domed shape and due to ligaments and joint capsules, the mobility is limited. The intercarpal (Articulationes intercarpales) denote the articulated joints of the carpal bones of a series with each other.

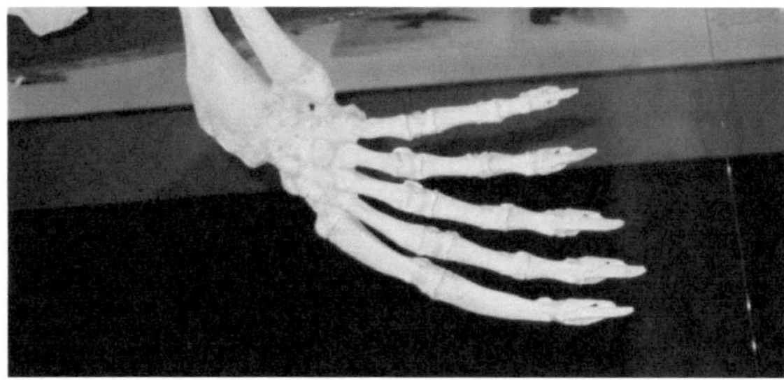

Illustration 9: Preparation of a sea lion hand. Natural History Museum Berlin, Mi. Dienst, 2013.

They are so-called loose joints (amphiarthroses), the stiffened by numerous strip tensions and can hardly move. The carpometacarpal joints indicate the connection of the distal carpal bones with the second to fifth metacarpals (Articulationes carpometacarpales). In humans, they are not directly associated with the wrist, but always expected in animals to pastern.

Offshore back into the water
The skeletons of the forelimbs of whales (Cetacaea) and cetaceans are wearing the typical structure of the mammalian hand. The upper arm is compact, fibula and Tabia are flattened. The fingers are of different lengths, carrying 4 to 12 segments and can even vary within an individual. In some species the metacarpals are not trained at all. The differentiation and morphogenesis process of the hands of whales and cetaceans may extend into adulthood depending on the type. Thus we find in the collections of a heterogeneous picture of whale hand before: of cartilaginous elements to an ossified center around younger, up to completely ossified systems and fused upper and lower arm in old animals. The modern whales evolved before 30-40 million years ago. morphologists are today no longer so sure whether primitive hoofed Eocene the forefathers modern whales (Cetacaea) and cetaceans were. Latest studies suggest that carnivorous ungulates that resembled in shape wolves, more to hunt and more coastal waters, estuaries and the sea and sought out a aquatiate life assumptions. The earliest known Urwal (Pakicetus) lived about 53 million years. His skull has not (yet) have much in common with those of land animals. Gradually changes in the course of a unique evolution campaign, the skeleton of the mammal, and it developed a streamlined body contours, the loss of his hair-dress is accompanied by the formation of the insulating and flow elastic layer of fat (blubber), the forelimb formed to fins (Flippers) to, hindlimb and pelvic girdle recede. Fluke and Finn are reinventions in the

phylogeny of the marine mammals. The last (protozoan-) whale disappear before about 30 million years ago and very quickly the modern whales develop with the two sub-systems of baleen and toothed whales. The basic kinematics of vertebrate skeletons we shall understand in simplified models. The graph Fig.10 tries a first schematization of complex transmission of vertebrate metacarpal. With an abstract point of view can be extracted basic gear assemblies.

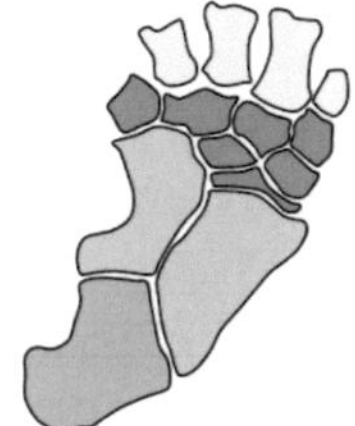  

Illustration10: Schematization of marine mammals-hand: Metacarpala (yellow); Centralia, carpals (red); Fibula, Tabia (green); Fermur (gray), on the left. Syntactic elements and derivable joint principles (right).

But what we know at all about under load deforming structures? Materials, construction methods, topology, shape and dimensions are by a designer by Optimal criteria (minimum, maximum) evaluate and select: function and strength, shape retention, safety, durability, no claims or damage-tolerance and compatibility with the surrounding environment, for example, Environmental concerns. In marine vessels the resilience criterion should be weighted in the future, the question of robustness and adaptability of a technical (artificial) system. The principle of engineers and designers, dimensioned construction in all parts makes sense only if the optimization concerns are classified as fully satisfied (or completely disregarded). From this paradigm, we will hereinafter (controlled) vary. The essential in fluid mechanically active components detection of static and dynamic strength, material control and manufacturing precision will be of course also require a CARPO fin, but in the end it is still a "the flow forces on approved" component, a structure with which we no gained experience and of which we know very little. The flow patterns, the physical impact is not suffered from a component with "intelligent mechanics (i-mech)", but - in a way, parasitic - used to induce a controlled contour, shape and functional transformation of the component. This is certainly innovative, future-proof but with the experience we still lack of such a component property, the right semantics; this is certainly a disadvantage. Consider the following remarks as an experiment. Elastic components which mechanically behave orthodox, deform in the direction of the acting force action. The shape orthodox deformed tensile structures is convex deformed against the non-impacted form. In some cases, the fluid-mechanical properties of such a deformed guidance and control airfoil can be advantageous and desirable, for example, as an overload protection. More commonly, the design intent is on a self-reinforcing effect. In these cases, the elastic members of the underwater hull should have a non-Orthodox impingement deformation behavior. The shape of a non-orthodox deformed wing is concave deformed against the non-impacted form. The wing deformation is initially independent of the flow respects a convex or concave deformed wing and the fluid mechanical properties of their (deformed) to consider profile contours.

As a rule, that phase of industrial product development (PE) accompanying, ever more refined and ultimately detail elaborated design (CAD), which the assumptions of the simulation software (data formats) and the operating and boundary conditions of a most realistic scenario (exists stresses, clamping conditions, etc.) in an FEM calculation program (finite element method, FEM) is transmitted. There are several fundamental and mutually distinct approaches in the engineering field. The calculation is performed for different-union (but usually rather late) stages of product development. Work results are statements about the strength and deformation behavior of the component and the voltage distribution in the interior of the structure. The information flow back into the design and development process. The usefulness of the information depends on the model preparation.

Bibliography, additional literature, patents and Internet links

[Albe-09]   Alben, S. (2009) On the swimming of a flexible body in a
             vortex street.   in J. Fluid Mech. (2009), vol. 635, pp. 27–45.
             Cambridge University Press 2009
[Albe-06]   Alben, S., Madden, P.G., Lauder, V.L. (2006) The mechanics of
             active fin-shape control in ray finned fishes. Journal of the Royal
             Society. Interface Vol.: 2007/4, S. 243-256.

[Ande-99]   Anderson, J.M. (1999) NEAR-BODY FLOW DYNAMICS IN SWIMMING FISH, The Journal of Experimental Biology 202, 2303–2327 (1999)

[Antm-05]   Antman S.S. (2005) Nonlinear problems of elasticity. 2nd edn. Springer; New York, NY.

[Batc-67]   Batchelor G.K. (1967) An introduction to fluid dynamics, 1st edn. Cambridge University Press; Cambridge, UK.

[Bann-02]   Bannasch, Rudolph. (2002) Vorbild Natur. In: design report 9/02, S.20ff. Blue.C Verlag Stuttgart.

[Bapp-99]   Bappert, R. Bionik, (1999) Zukunftstechnik lernt von der Natur. SiemensForum München/Berlin und Landesmuseum für Technik und Arbeit (Herausgeber).

[Barg-11]   Bagaric, B. (2011). Modellierung, Simulation und Parametrisierung eines virtuellen Strömungskanals mit dem Programmsystem FS-Flow. Untersuchung typischer Szenarien endlicher Traglügel. Bachelorarbeit a.d. BeuthHS Berlin (082011).

[Bech-93]   Bechert, D.W.: Verminderung des Strömungswiderstandes durch bionische Oberflächen. In: VDI-Technologieanalyse Bionik, S. 74 – 77. VDI-Technologiezentrum Düsseldorf 1993.

[Bech-97]   Bechert, D.W., Biological Surfaces and their Technological Application. 28th AIAA Fluid Dynamics Conference: 1997

[Curr-25]   Curry, M. (1925) Die Aerodynamik des Segels und die Kunst des Regatta-Segelns. Diessen vor München: Jos. C. Huber, 1925.

Floc-09]   France Floch,F. Laurens, J.M. (2009) Comparison of hydrodynamics performances of a porpoising foil and a propeller. in: First International Symposium on Marine Propulsors smp'09, Trondheim, Norway, June 2009

[Gopa-94]   Gopalkrishnan, R.(1994) Active vorticity control in a shear flow using a flapping foil. in J. Fluid Mech. (1994), vol. 274, pp. 1-21 Cambridge University Press.

[Hild-01]   Hildebrand, M., Goslow, G.E., (2001) Vergleichende und funktionelle Anatomie der Wirbeltiere. Springer Verlag Berlin, N.Y.

[Liao-03]   Liao, J.C.; Beal, D.; Lauder, G.; Triantayllou, M. (2003): Fish Exploting Vortices Decrease Muscle Activty, In: Science 2003, S. 1566-1569. AAAS.

[Liao-06]   Liao, J.C.; Passive propulsion in vortex wakes. in J. Fluid Mech. (2006), vol. 549, pp. 385–402. c_ 2006 Cambridge University Press

[Kreb-08-2] Krebber, B.: "i-mech". Untersuchung der intelligenten Mechanik von Fischflossen mit Hilfe von FSI- Simulation. Forschungsbericht der Technischen Fachhochschule Berlin 2007/08

[Kreb-08-1] Krebber, B., H.-D. Kleinschrodt und K. Hochkirch: (2008) Fluid-Struktur-Simulation zur Untersuchung intelligenter Mechanik von Fischflossen. ANSYS Conference & 26. CADFEM Users´ Meeting,

[McCu-70] McCutchen C.W. (1970) The trout tail fin, a self-cambering hydrofoil. J. Biomech. 1970/3, S. 271–281.

[Nach-98] Nachtigall, W.: Bionik. Grundlagen und Beispiele für Ingenieure und Naturwissenschaftler. Springer-Verlag, Berlin-Heidelberg-New York 1998.

[Nach-00] Nachtigall, W.; Blüchel, K. Das große Buch der Bionik. Stuttgart: Deutsche Verlags Anstalt: 2000.

[Mirs-05] Mirtsch, F.; Dienst, M.: FlowBow-Artifizielle adaptive Strömungskörper nach dem Vorbild der Natur. In: Forschungsbericht der Technischen Fachhochschule Berlin 2005

[PaBe-93] Pahl. G.; Beitz, W.: Konstruktionslehre, 3.Auflage. Berlin-Heidelberg- New York-London-Paris-Tokio: Springer 1993

[Pfeif-07] Pfeiffer,Rolf; Bongard, Josh (2007): How the body shapes the way we think, The MIT Press

[Read-02] D.A. Read (2002) Forces on oscillating foils for propulsion and maneuvering, in Journal of Fluids and Structures 17 (2003) 163–183 Cambridge University Press

[Rech-94] Rechenberg, Ingo, (1994) Evolutionsstrategie. Frommann Holzboog Verlag Stuttgart- Bad Cannstatt.

[Sege-87] Segel L.A. (1987) Mathematics applied to continuum mechanics. 1st edn. Dover Publications; New York, NY.

[Siew-10] Siewert, M; Kleinschrodt, H-D; Krebber, B; Dienst, Mi. (2010) FSI-Analyse auto-adaptiver Profile für Strömungsleitflächen. In: Tagungsband, ANSYS Conference & 28th CADFEM Users' Meeting Aachen 2010.

[Siew-11] Siewert, M; Kleinschrodt, H-D.(2011) Bionical Morphological Computation. In: Nachhaltige Forschung in Wachstumsbereichen Bd.1, Logos Verlag Berlin.

[Stre-96] Streitlien, K. (1996) Efficient foil propulsion through vortex control, Aiaa Journal - AIAA J , vol. 34, no. 11, pp. 2315-2319, 1996

[Tria-95]   Triantafyllou, M. (1995): Effizienter Flossenantrieb für Schwimmroboter, Spektrum der Wissenschaft 08-1995, S. 66–73, Wiss. Verlagsges. mbH, Heidelberg 1995.

[Tria-95]   Triantafyllou, M., Triantafyllou, G. (1996): An Efficient Swimming Machine. Scientific American, March 1996. p.64-70.

[Tria-02]   Triantafyllou, M. (2002) Vorticity Control in Fish-like Propulsion and Maneuvering, INTEGR. COMP. BIOL., 42:1026–1031 (2002)

[Die12-US]   Dienst, Mi. (2012) COMPONENTS DESIGNED TO BE LOAD-ADAPTIVE. US-Pat. 13/517,181 (based on PCT/DE2010/075164, 19062012).

[Die12-WO]   Dienst, Mi. (2012) COMPONENTS DESIGNED TO BE LOAD-ADAPTIVE. WO: PCT/DE2010/075164 (based on PCT/DE2010/075164, 19062012). IPC: B63H (2012.01)

[Die12-EU]   Dienst, Mi. (2012) COMPONENTS DESIGNED TO BE LOAD-ADAPTIVE. EU-Pat. 10809144.8 (based on PCT/DE2010/075164,).

[Die12-DE]   Dienst, Mi. (2012) Belastungsadaptiv ausgebildete Bauteile. Dt. Patent PTC/DE2010/075164, EP: 10809144.8, Offenlegung. 22062011

Graphics and Images (originals)

Abb. 1: Testobjekte aus Pappe. Mi. Dienst (2016)

Abb.2: Präparat einer „modernen" Delfinhand. Naturkundemuseum Berlin, Mi. Dienst (2013).

Abb.3 und Abb.4: Präparat einer „modernen" Delfinhand. Naturkundemuseum Berlin, Mi. Dienst (2013).

Abb.5: und Flossenmembran mit Flossenstrahlen (schematische Darstellung) Steg S, Halbtube H, Membran M, Fuge F und Inlet I.

Abb.6: Das nichtorthodoxe Beaufschagungs-Bewegungs-Gebaren einer Makrelenfinne. Mi. Dienst (2013).

Abb.7: Organisation und Prinzip der Entwicklung des Extremitätenskeletts der Tetrapoden. Schematisch nach Schubin: ontogenetische Entwicklungsachse gepunktet gezeichnet. Mi. Dienst (2016)

Abb.8: Präparat eines Saurierfußes (Kentrosaurus). Naturkundemuseum Berlin, Mi. Dienst (2013).

Abb.9: Präparat einer Seelöwenhand. Naturkundemuseum Berlin, Mi. Dienst (2013).

Abb.10: Schematisierung der Meeressäuger-Hand: Metacarpala (gelb); Centralia, Carpalia (rot); Fibula, Tabia (grün); Fermur (grau), links im Bild. Syntaktische Elemente und ableitbare Gelenkprinzipien (rechts im Bild). Mi. Dienst (2016).